SCIENCE FACTORY

FORCES & SIMPLE MACHINES

JON RICHARDS

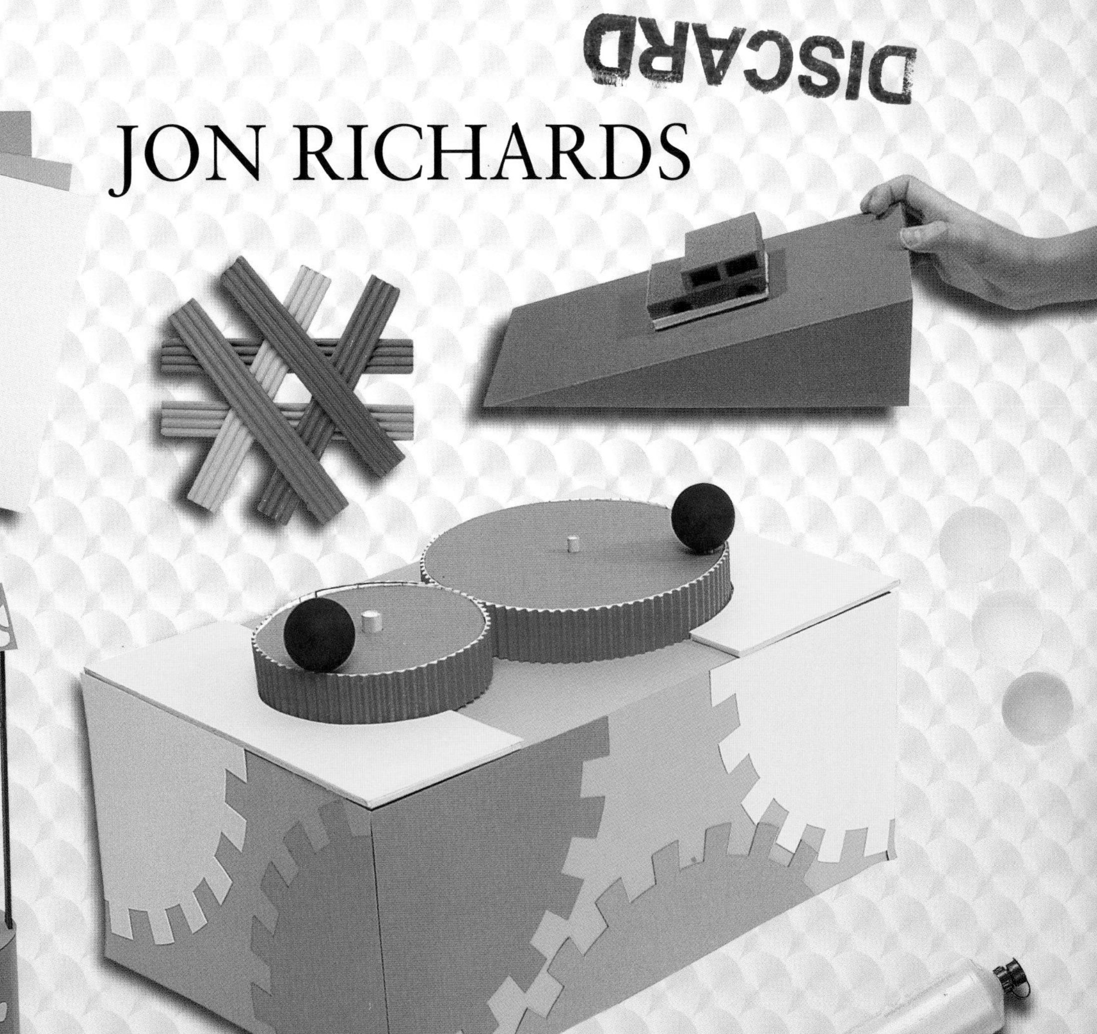

PowerKiDS press

New York

Published in 2008 by The Rosen Publishing Group, Inc.
29 East 21st Street, New York, NY 10010

Editor:
Kathy Gemmell

Design:
David West Books

Designer:
Jennifer Skelly

Illustrator:
Ian Moores

Consultant:
Steve Parker

Photographer:
By Roger Vlitos

Library of Congress Cataloging-in-Publication Data

Richards, Jon, 1970–
Forces & simple machines / Jon Richards.
p. cm. — (Science factory)
Includes index.
ISBN-13: 978-1-4042-3908-1 (library binding)
ISBN-10: 1-4042-3908-1 (library binding)
1. Simple machines—Juvenile literature. I. Title.
TJ147.R525 2008
621.8078—dc22
2007016596

Manufactured in the United States of America

INTRODUCTION

Simple machines are all around us all the time – even parts of our bodies work like simple machines. Machines are used to overcome forces, such as friction and gravity. However complicated they may seem, most machines are made up of just a few central parts, including levers, wheels, and pulleys. A seesaw is a type of lever. Read on and discover a host of projects to teach you more about forces and simple machines.

CONTENTS

YOUR FACTORY

BEFORE YOU START any of the projects, it is important that you learn a few simple rules about the care of your science factory.

- Always keep your hands and the work surfaces clean. Dirt can damage results and ruin a project.

- Read the instructions carefully before you start each project.

- Make sure you have all the equipment you need for the project (see checklist opposite).

- If you haven't got the right piece of equipment, then improvise. For example, a strip of cardboard rolled into a ring will do just as well as a wide roll of tape.

- Don't be afraid to make mistakes. Just start again – patience is very important!

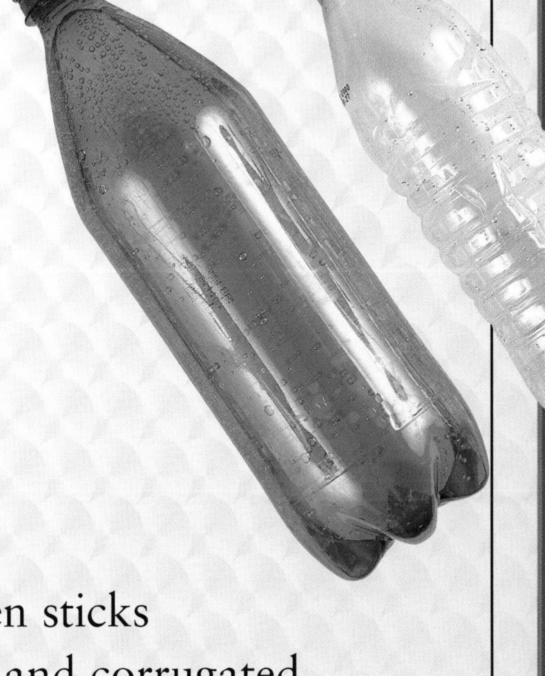

Equipment checklist:

- Modeling clay
- Drinking straws (wide and narrow)
- Marbles and table-tennis balls
- Corks
- Tape and glue
- Cardboard boxes
- Paper fasteners and thumbtacks
- Toy car
- Toothpicks, matchsticks, and wooden sticks
- Colored cardboard, stiff cardboard, and corrugated cardboard
- Wide and narrow elastic bands

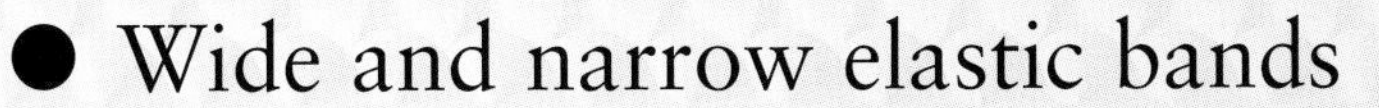

- Scissors and stapler
- Cotton thread and thread spools
- Paints
- Candle
- Four round sponges
- String
- Wide roll of tape
- Compass

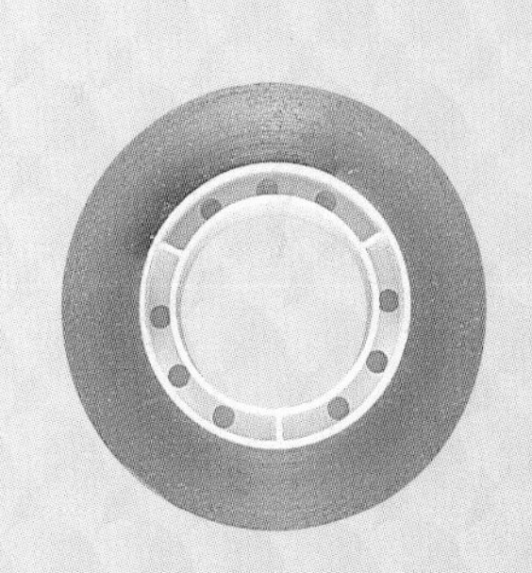

WARNING:
Some of the projects in this book need the help of an adult. Always ask a grown-up to give you a hand when you are using scissors or tools like staplers.

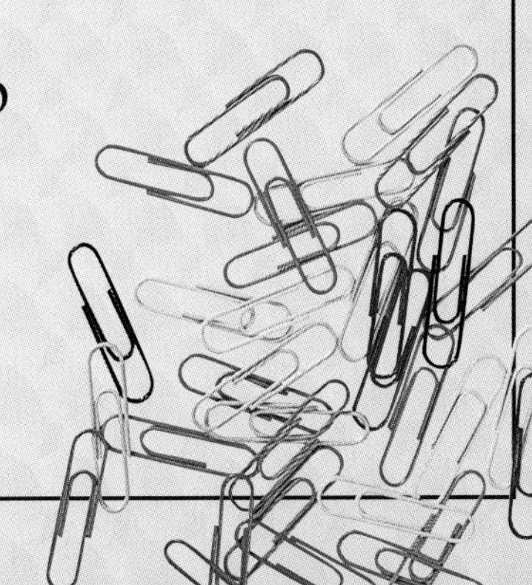

LEVERS

WHAT YOU NEED
Colored cardboard
Drinking straws
Marble
Cork, Tape
Scissors
Thumbtacks
Modeling clay

MACHINES ARE DEVICES that make work easier. One of the most common machines is a lever. A lever is simply an arm that can move an object with the help of a pivot. We use levers all the time every day. A spade, a bottle opener, and even our arms and legs, are levers. This project shows you how to build a contraption that uses three different types of lever. Ask an adult to help you with all the steps.

FULL FORCE

1 For the first lever, cut out a pointing hand and tape it to a length of straw, as shown here. Use modeling clay to attach this to one end of a strip of cardboard. Tape a short length of straw to the bottom of the strip (A).

2 For the second lever, make a tall box out of cardboard so it is the same height as the pointing hand. Fold another strip of cardboard to make a channel for the marble to run along.

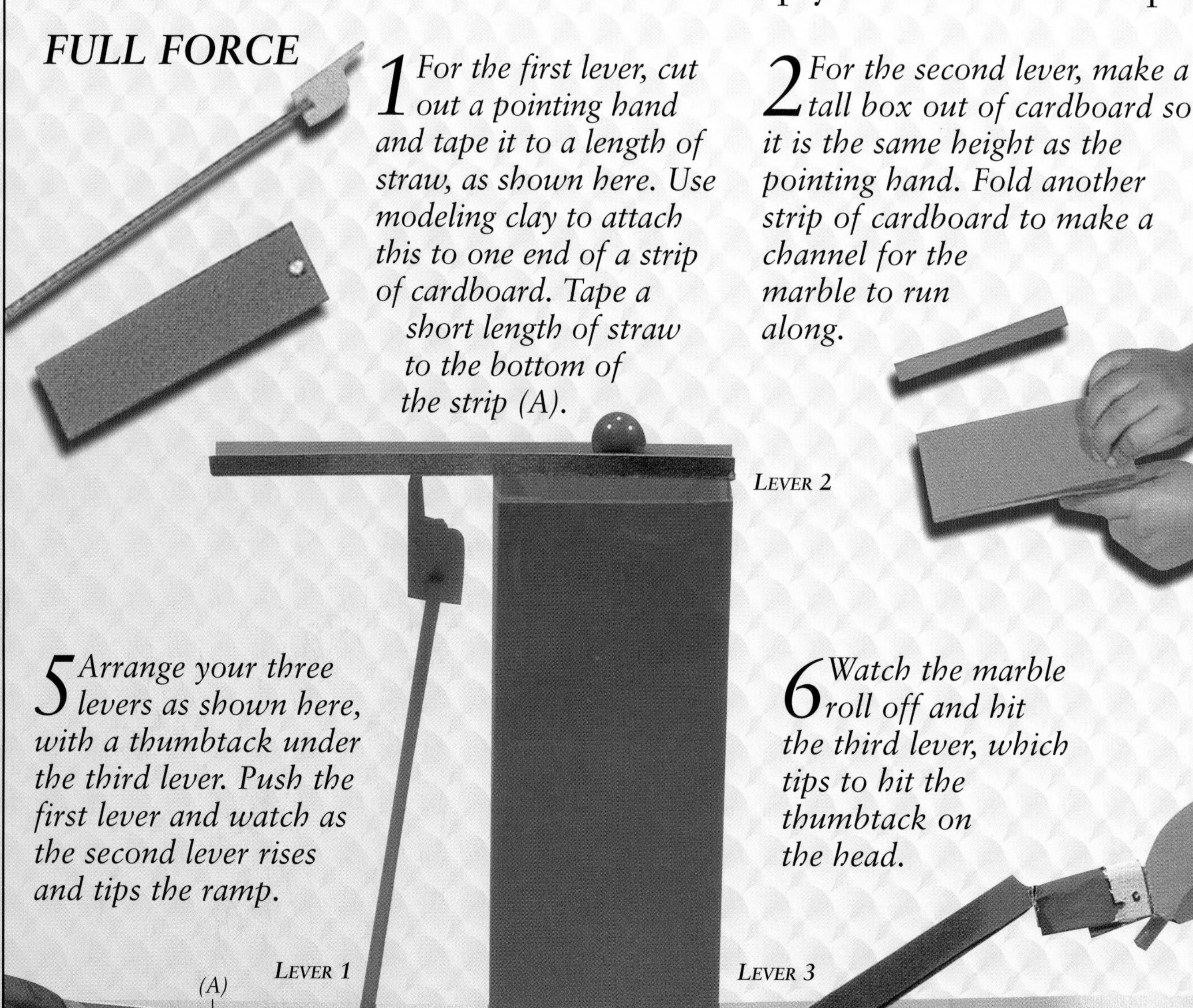

5 Arrange your three levers as shown here, with a thumbtack under the third lever. Push the first lever and watch as the second lever rises and tips the ramp.

6 Watch the marble roll off and hit the third lever, which tips to hit the thumbtack on the head.

3 *Tape one end of the cardboard channel to one side of the top of the tall box. Make sure the other end of the channel can move up and down easily.*

4 *For the third lever, make a hammer using a piece of straw, a thumbtack, and a cork, as shown. Cut out another pointing hand, tape it to the hammer, and attach it to another cardboard strip with clay. Now form a pyramid with three short lengths of straw and tape it to the bottom of the strip (B).*

WHICH LEVER?

What types of lever are shown here? Look for other levers that are used at home. See if you can figure out what types of levers they are.

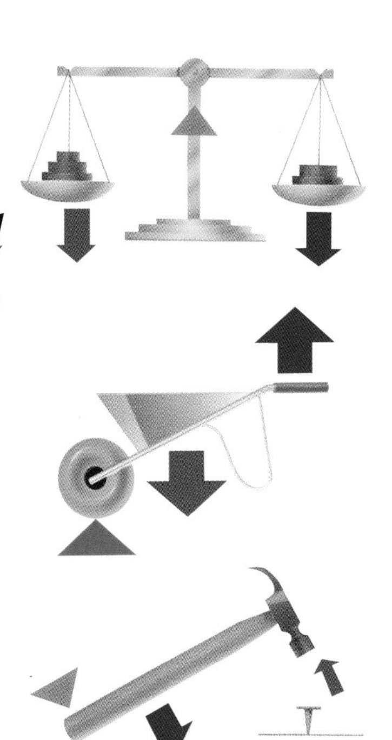

WHY IT WORKS

Your machine uses all three levers. The type of lever depends on where you find the pivot, the load, and the force needed to move it. A first-class lever (1) has the pivot between the force and the load. A second-class lever (2) has the load between the pivot and the force. A third-class lever (3) has the force between the load and the pivot.

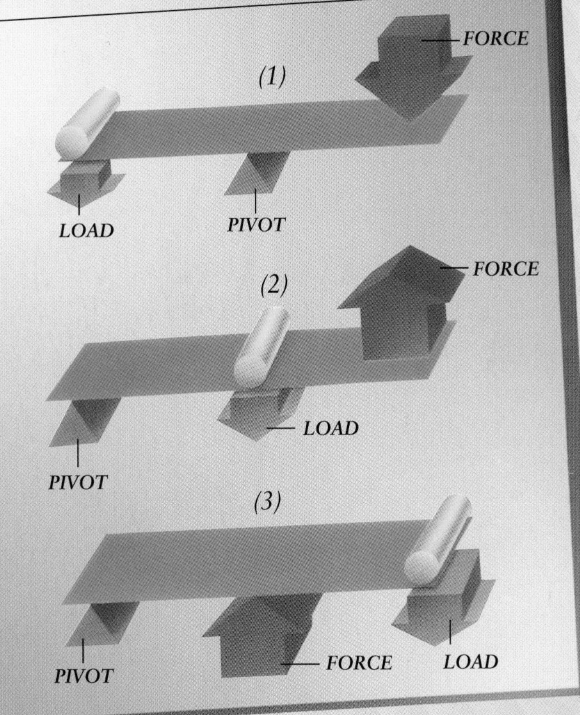

WORKING IN PAIRS

WHAT YOU NEED
Colored cardboard
Cardboard boxes
Paper fasteners
Thumbtacks
Glue
Matchstick
Cotton thread
Elastic band
Drinking straw
Stapler
Scissors

SOMETIMES ONE LEVER is not enough to do a job. Scissors use two levers against each other to cut through an object, just as a pair of pliers uses two to grip something. You even find levers working in pairs on your own body. Pick up a pencil and you will see your fingers and thumb working against each other to grip the pencil. Build your own grabber in this project to see levers working in pairs.

LEVERS IN PAIRS

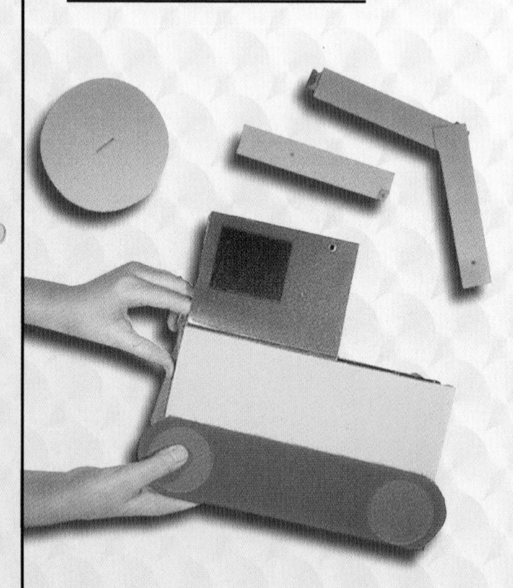

1 Ask an adult to cut out three strips and a circle of cardboard with a slit in it, as shown. For the grabber's arm, glue two strips together as shown and attach on the last strip with a paper fastener. For the base, glue together two boxes and decorate them with colored cardboard.

2 Cut out two cardboard jaws. Attach the bottom set to the top set with a paper fastener. Glue the top set to the arm so the bottom set can still move. Glue a matchstick to the top set, as shown. Tie a long piece of thread to the matchstick.

3 Run the thread along the arm, through two lengths of straw, glued on as shown. Staple the elastic band to the arm and the bottom jaws to keep the jaws open.

4 Bend a length of straw and push it through the body. Tape the end of the thread to the straw to make a handle. Make sure it can turn.

WHY IT WORKS

As you wind the handle, you reel in the thread, shortening it. This exerts a force on the arm. The parts of the arm act as levers, straightening around the pivot made by the paper fastener. The force is transmitted along the arm to the lower jaws, which also act as a lever.

LOOK FOR LEVERS

Look around your home to see other levers working in pairs. You will find them in scissors or tweezers. See if you can work out where the pivot is in each case.

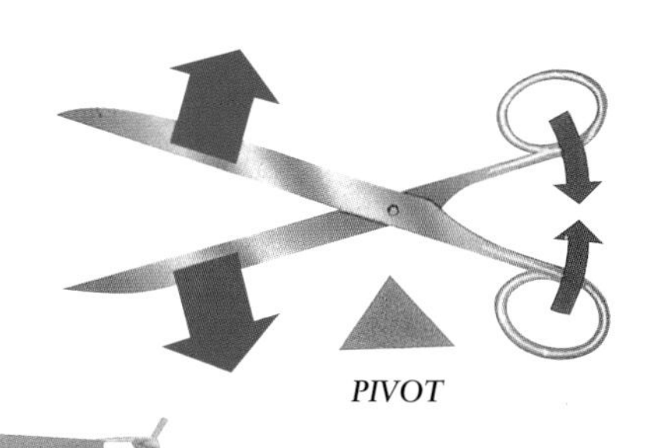

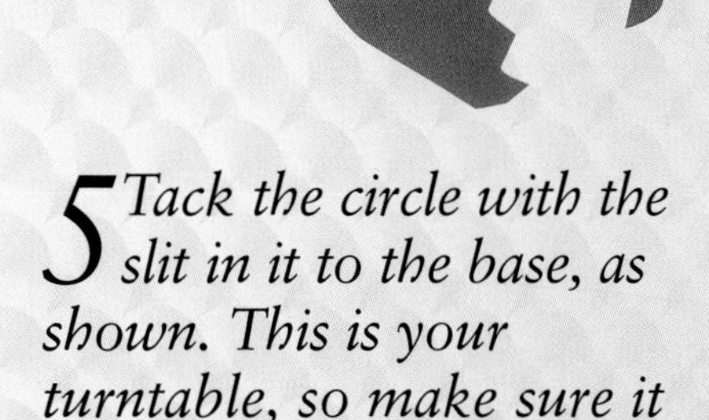

5 Tack the circle with the slit in it to the base, as shown. This is your turntable, so make sure it can move.

6 Slot the grabber's arm into the slit in the turntable. Turn the handle and watch as the arm straightens and the jaws close.

*F*RICTION

WHAT YOU NEED
Stiff cardboard
Wide and narrow drinking straws
Elastic band
Paper fasteners
Paints
Toothpick
Glue
Scissors

FOR MACHINES TO DO WORK, they must overcome certain forces. One of these forces is friction. Friction occurs when two things rub against each other. When you ride your bike, there is friction between the bike and the road and also between the bike, yourself, and the air. This has the effect of slowing you down. This project lets you see how friction slows some objects more than others.

WARMING UP

Rub your hands and they start to feel warm. As they rub against each other, the roughness of your hands creates friction. Energy lost in overcoming the friction shows itself as heat, and your hands feel warm.

WHY IT WORKS

Friction occurs between the objects and the strip, making the objects slow and stop. The amount of friction is affected by the weight, the smoothness, and the area of the object in contact with the strip. Heavy, rough, or large objects cause more friction and do not travel as far as light, smooth, or small objects.

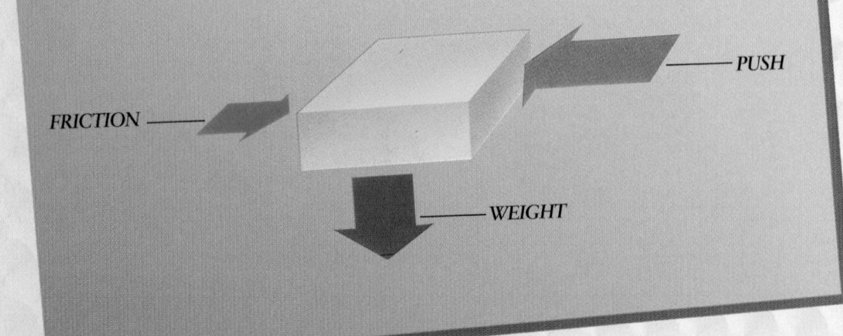

MOVING THINGS

1 Paint stripes across a long strip of cardboard. Cut and glue a short length of wide drinking straw in the middle of one end, as shown.

2 Make a T-shaped hammer by sticking two narrow straws together. Slide this into the wide straw stuck to the cardboard strip.

3 Push the toothpick sideways through the narrow straw. Push fasteners through the strip on either side of the hammer, as shown. Cut open an elastic band. Stretch it between the fasteners, hooking it behind the narrow straw.

4 Pull the hammer back and place an object in front of it. Release the hammer and see how far it manages to push the object along the strip.

5 Try several different objects and see how far your hammer pushes them along the strip of cardboard. Which objects travel the farthest and which travel the least?

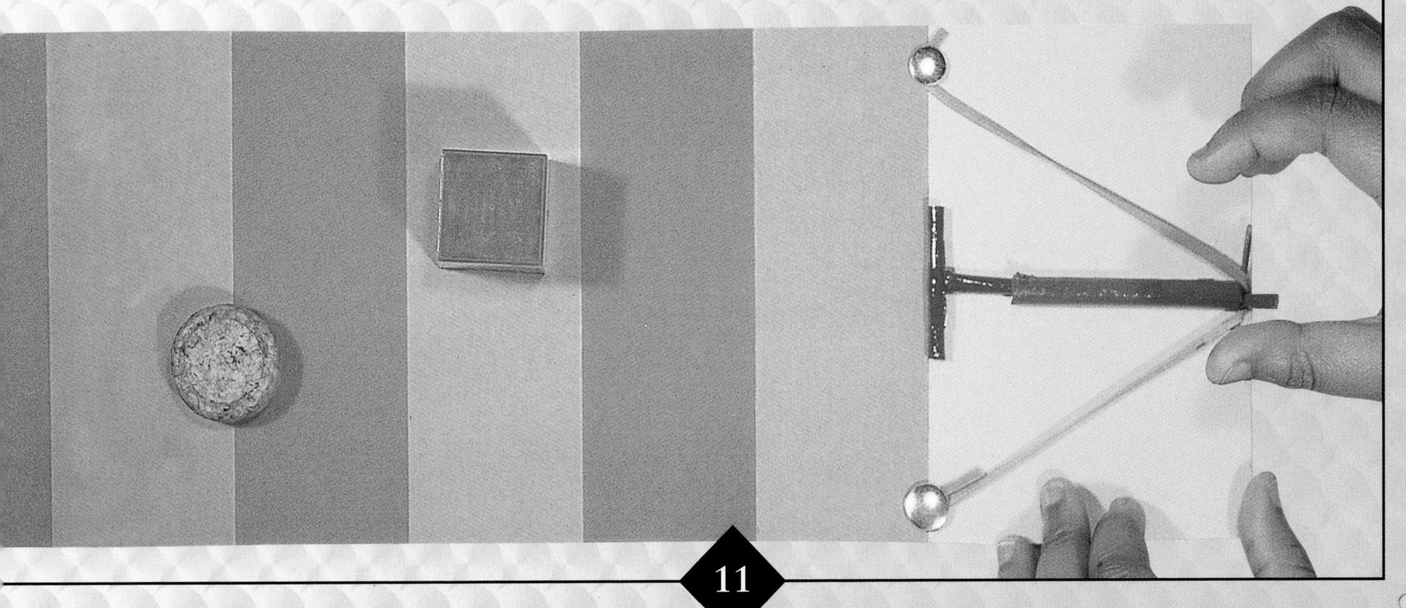

CLIMB THE HILL

WHAT YOU NEED
Stiff cardboard
Candle
Elastic bands
Toothpick
Thread spool
Tape
Drinking straw
Ruler
Scissors

THE LAST PROJECT SHOWED HOW friction is one of the forces involved when dealing with machines, and how it can slow objects down. However, friction does have its uses. Without friction you would not be able to stop when you ride your bike. You need the friction between the brake pads and your bike's wheels to slow you down. This project shows you another way that friction can be useful.

CLIMBING TANK

1 Ask an adult to cut a short length off a candle and bore a hole through it. Push an elastic band through this hole and hold it in place with a toothpick, as shown.

2 Push the other end of the elastic band through the thread spool and secure it in place with tape.

3 Make a ramp from stiff cardboard (see page 24). Wind the toothpick as tight as it will go. Place the spool at the bottom of the ramp and see if it climbs up.

4 Now attach some elastic band tires around the thread spool. Repeat the process and you will find that the spool can climb the ramp better this time.

5 Cut out a body shape for your tank from cardboard. Tape on a drinking straw for a gun. Fold the cardboard tank over the spool, as shown above.

WHY IT WORKS

The tank uses the friction between the elastic bands and the slope to give it the grip to climb. Without these tires, there would not be enough friction to create enough grip for the tank to climb the slope. On the other hand, the candle wax reduces the friction between the toothpick and the thread spool, so the spool can spin more steadily as the elastic band unwinds. Without the candle wax, there might be too much friction for the spool to spin.

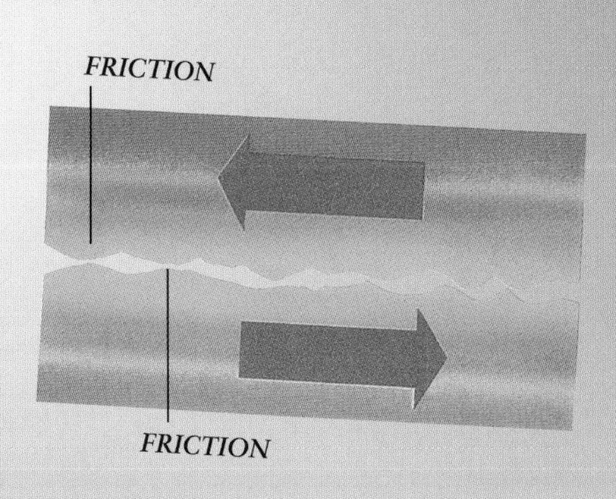

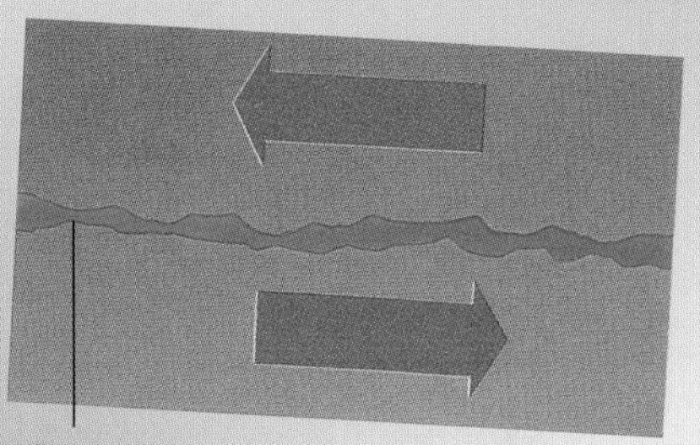

CANDLE WAX REDUCES FRICTION

6 Place your tank at the bottom of the ramp, wind the toothpick, and watch as your tank climbs the slope.

STEEPER SLOPE

How steep a slope can your tank climb? Try different elastic band sizes. Do thick or thin ones give the best grip?

SMOOTH RUNNING

WHAT YOU NEED
Stiff cardboard
Drinking straws
Marbles
Ruler
Scissors
Glue

YOU SAW IN THE LAST project how friction can be a useful force, providing machines with the grip to climb a slope. However, friction is not always useful. Sometimes it gets in the way, especially if you are trying to get from one place to another as quickly and as easily as possible. Friction makes this harder and means you have to use more energy when traveling. See how the effects of friction can be reduced with this project.

SLIP SLIDING

1 Make two small boxes and glue one on top of the other to make a car shape, as shown here. Make another shallow box that is no deeper than the width of a straw.

2 Place the car body on a ramp (see page 24 for how to make a ramp). You will see that it does not slide down. Tilt the ramp until it does slide. You will have to tilt the ramp a long way.

3 Ask an adult to cut lengths of drinking straw so they fit inside the shallow box you made.

4 Place the shallow box upside down on the ramp and put the car body on top. Tilt the ramp until the car begins to move. You will need to tilt the ramp a lot less than when the car did not have the straws.

5 Make two narrow trays, big enough to hold a line of marbles.

ROUGHING IT UP

Try covering your ramp with different materials, such as sandpaper or a piece of carpet. Does your car move down the slope as easily this time?

6 Carefully place the marble trays on the ramp and place the car body on top. You should not have to tilt the ramp much before the car moves down the slope.

WHY IT WORKS

The car body on its own does not slide because the friction between itself and the ramp is too great. Using straws or marbles as wheels reduces friction because their round shape reduces the area of contact between the car and the ramp. Marbles reduce the friction more because there is less contact between them and the ramp.

WHEEL AND AXLE

WHAT YOU NEED
Colored cardboard
Stiff cardboard
Four round sponges
Scissors
Wooden sticks
Tape
Glue
Strong drinking straws

IN THE LAST PROJECT, you used marbles and straws as wheels. Wheels are good for moving objects because their round shape reduces the area of contact between surfaces and so reduces friction. However, a vehicle needs more than rigid wheels if it is to be useful. To get around corners, these wheels need to turn. Early carts used a simple axle to achieve this. See how one works in this project.

STEER CLEAR

1 Ask an adult to cut out a steering wheel and two car shapes from colored cardboard, as shown, for the sides of the car. Leave spaces for the wheels, but make sure you leave flaps hanging from the back wheel arches, as shown. Glue more cardboard on for the windows.

GOING BACKWARDS

Try pushing your car in reverse and steering. You will notice that the car follows a different path when the wheels are steering from the back.

2 Ask an adult to cut a wide strip of cardboard for the car body. Tape the body to the sides, under the car. Turn it upside down to do this. Glue on windows, headlights, and a radiator flap at the front.

3 Make a steering column and axle by sticking two straws together with lots of tape, as shown.

4 Turn the car over. At the front end, stick a strip of stiff cardboard between the sides of the car, with another strip on top, as shown. Ask an adult to push the T-shaped straws through both strips and through the body of the car.

5 Make the wheels by sticking a circle of cardboard on each sponge. For the front wheels, push a stick into one sponge. Thread the stick through the T-shaped part of the axle you made with straws. Then stick another wheel on the other end.

WHY IT WORKS

By turning the steering wheel, you are moving the axle and the wheels away from the straight-on path of the car. Because the car then has to follow the new path that the wheels are pointing in, the car turns a corner.

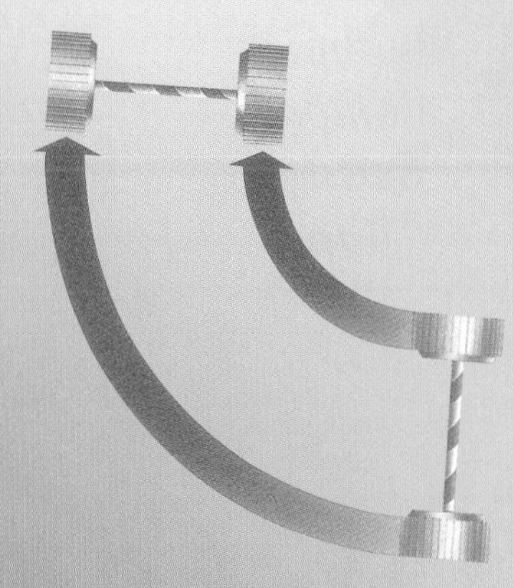

6 For the back wheels, thread a stick through the hanging flaps and stick a sponge on each end. Tape on your steering wheel, and off you go!

*L*IFTING A LOAD

WHAT YOU NEED
Stiff and colored cardboard
Drinking straws
Wooden stick
String
Glue
Tape
Scissors

THE PREVIOUS PROJECTS SHOWED you how wheels can be useful in machines to reduce friction and conserve energy. However, wheels can also be used for other important tasks. A machine called a winch is basically a wheel which, when turned, winds up a length of rope to lift something. You can see this in action with a mini water well which you can make in this project.

WHY IT WORKS

The shape of the handle means that your hand moves a greater distance than the distance turned by the straw. This reduces the amount of work you do, but means you are doing the work over a longer distance.

Learn about this using slopes in the next projects.

WINDING POWERS

1 Ask an adult to cut two strips of stiff cardboard and to punch a hole in each of them, as shown. Cut out and decorate a roof shape from colored cardboard.

2 Ask an adult to cut out a card-board ring, as shown here, and to cut a slit on each side just wide enough to hold the cardboard strips.

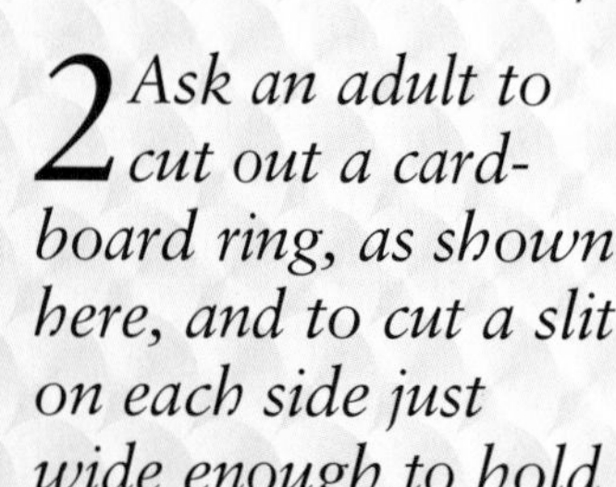

3 Ask an adult to thread a stick through a small piece of straw and two small circles of cardboard, as shown. Hold it in place with tape.

4 Roll and tape cardboard into a tube. Cut out a small circle of cardboard and tape it to the bottom for the bucket. Ask an adult to make small holes at the top, and thread string through them. Knot the ends.

5 Tie the bucket onto the small piece of straw, so that it will wind up, as shown. Push the stick through the holes in the cardboard strips. Bend a straw into a handle shape and thread it onto the end of the stick. Make sure it can turn easily. Tape on the roof shape you made earlier.

MAKING IT EASIER

Make the length of the handle longer. You will find it even easier to wind up the bucket. This is because your hand moves even farther than before, so the amount of work is less.

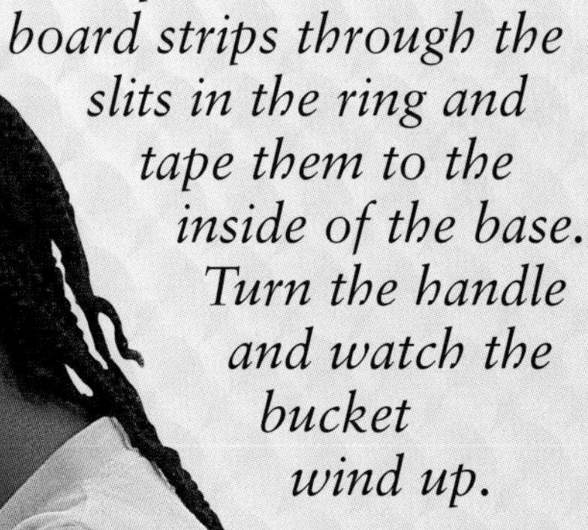

6 Make a well base in the same way you made the bucket, so that the ring fits on top. Slot the card-board strips through the slits in the ring and tape them to the inside of the base. Turn the handle and watch the bucket wind up.

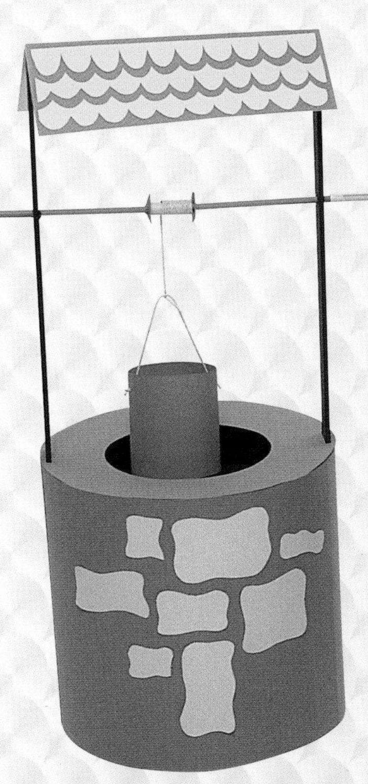

*R*OPES AND PULLEYS

WHAT YOU NEED
Colored cardboard
Glue, String
Drinking straws
Scissors
Modeling clay

THE LAST PROJECT SHOWED how winches and handles can reduce the level of work you do by increasing the distance over which it is done. Other machines used to reduce the level of work are pulleys. These are systems of wheels which, when ropes are wrapped around them, increase the distance greatly and so reduce the work level. You can build your own pulley system in this project.

HANGING AROUND

For another pulley, thread some string between two coat hangers. The more you thread the string between the hangers, the easier it is to lift a load. But you have to pull for longer each time, so the total work done is the same.

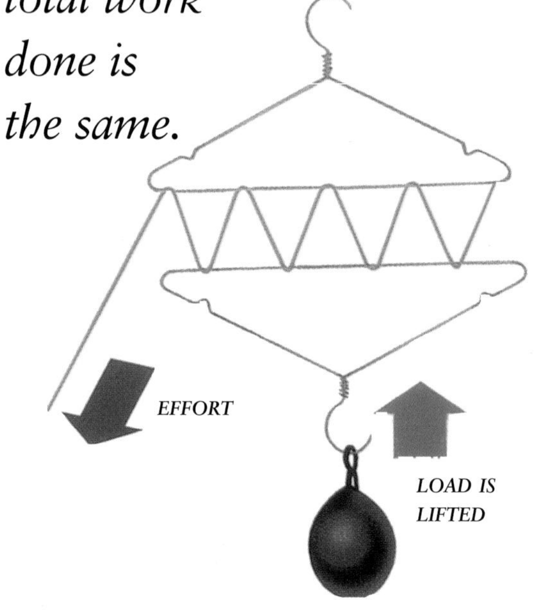

PULLING IT UP

1 Ask an adult to cut out the cardboard shapes shown here and to punch holes through the middle of the circles and on the other pieces of cardboard, as shown.

2 Glue the two circles of cardboard together, with a smaller circle of thick cardboard between them to make a thick wheel. Make another of these wheels.

WHY IT WORKS

By using two pulleys, you double the distance you have to pull with one pulley. This halves the work level and makes lifting the load easier. But you have to work for twice as long, so the total work done is the same.

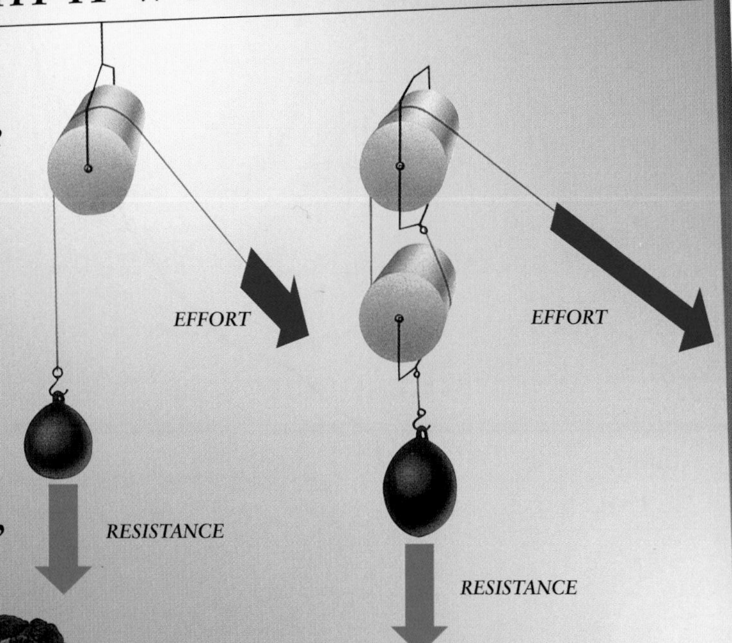

3 Sandwich each wheel between the cardboard shapes you cut out, as shown, and push a drinking straw through the holes in each to make your pulleys. Attach a handle to one pulley and a lump of modeling clay to the other.

4 Tie the string to the bottom of the pulley with the handle, thread it around the wheel of the other pulley, and then up over the wheel of the first pulley, as shown. Pull on this end of the pulley and watch as the load rises.

*U*P THE HILL

WHAT YOU NEED
Stiff, colored cardboard
Toy car
String
Drinking straws
Tape, Glue
Modeling clay
Ruler
Scissors

YOU MIGHT FIND walking up a slope hard work, but have you ever wondered how hard it would be to climb the same height as the slope, only vertically (straight up)? You may travel less distance, but the effort required is a lot greater. You can compare how much easier a slope makes it to climb a height in this project.

PULLING POWER

1 Ask an adult to help you make a ramp by cutting and folding a piece of cardboard as shown. Use the ruler and scissors to score a sharp fold line.

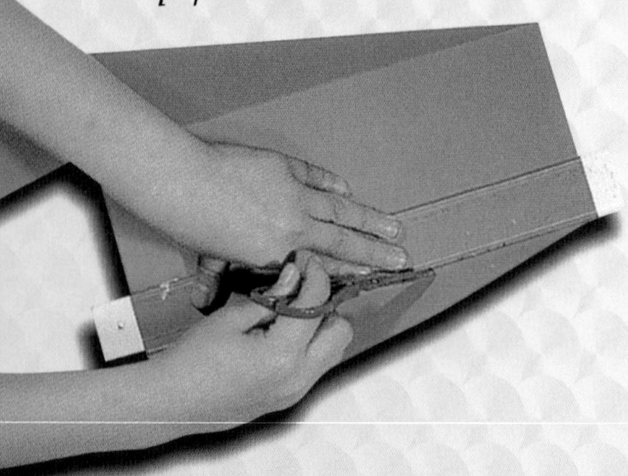

2 Tape the straws together to form a frame as shown. Glue on squares of cardboard to strengthen the frame, as shown opposite. Push a lump of modeling clay onto a piece of string, and cover it with cardboard to make a bucket. Tie the other end of the string to the toy car.

SCREWING IT UP

Screws are basically slopes that are wound around a central point. Test this by taking a triangular piece of paper and wrapping it around a pencil. Start at the shortest side of the triangle. The paper forms a screw pattern around the pencil.

START WRAPPING HERE

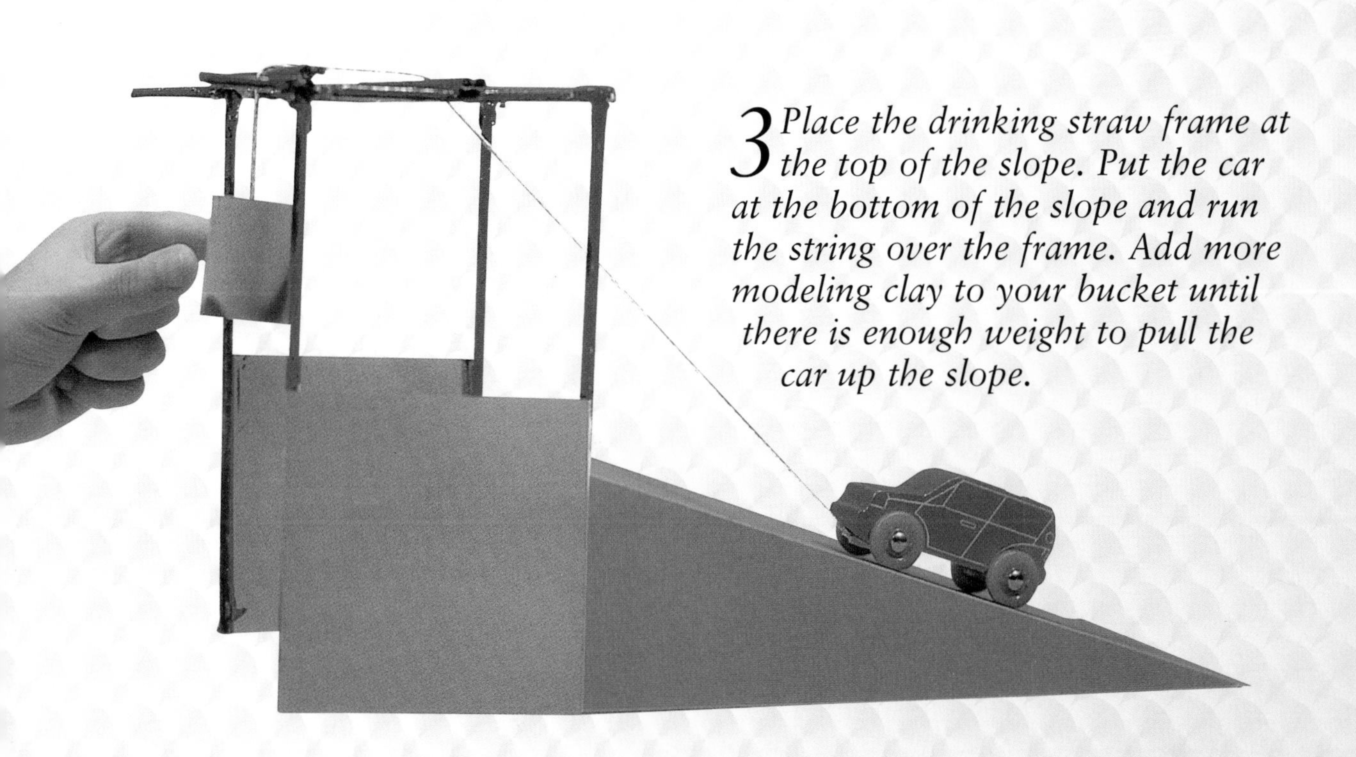

3 Place the drinking straw frame at the top of the slope. Put the car at the bottom of the slope and run the string over the frame. Add more modeling clay to your bucket until there is enough weight to pull the car up the slope.

4 Now place the car directly under the frame. Add more modeling clay to your bucket. You will find that much more weight is needed to raise the car vertically.

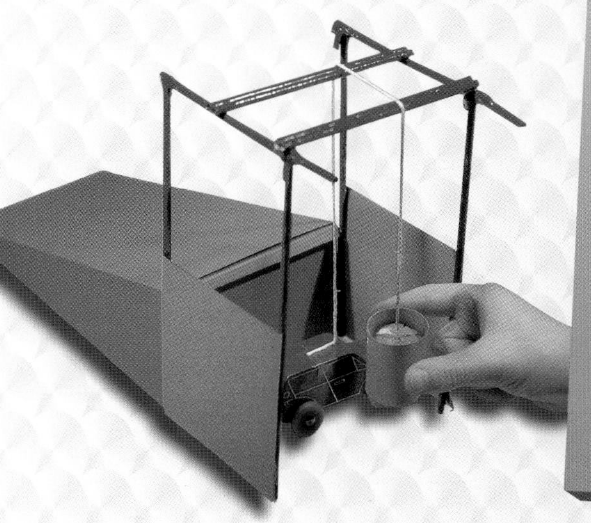

WHY IT WORKS

By moving the angle of work away from the vertical, a slope reduces the effects of gravity and reduces the level of work done. At the same time, however, it increases the distance the load has to move. So the total work done is the same.

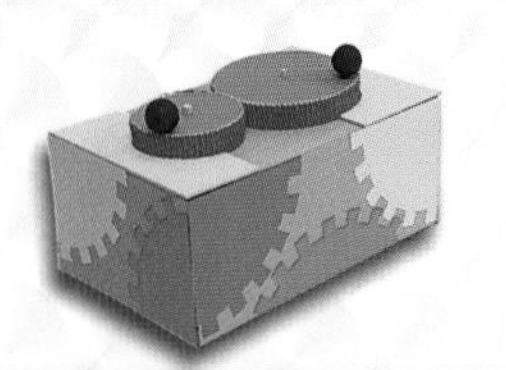

WHEELS AND COGS

WHAT YOU NEED
Colored cardboard
Corrugated cardboard, Glue
Table-tennis balls
Cardboard box
Wooden sticks
Scissors

IN OTHER PROJECTS IN THIS BOOK, you have seen how machines can be used to transmit power. One way of transmitting power is by using gears. The most common gear is the cog – a wheel with teeth. You may have seen cogs on a bicycle. Build some cogs and see the effects of using different sizes.

MAKING COGS

1 Ask an adult to cut out pairs of cardboard circles and thin strips from the corrugated cardboard. Glue the pairs of circles together with a strip of corrugated cardboard between them. Make sure that the ribbed side of the corrugated cardboard is facing out. These are your cogs.

2 Push a piece of stick through a hole in the center of each cog, so that it comes out the other side. Push a table-tennis ball stuck on a straw through a hole in the edge of each cog.

WHY IT WORKS

Because the ribs of each cog interlock, when you turn the first cog, the turning force is transmitted to the second cog, causing it to turn. If you turn the smaller cog one rotation, the larger wheel will not rotate as much. But if you turn the larger cog once, the smaller cog turns more than once.

3 Make holes in a cardboard box and push the wooden sticks through them. Ask an adult to help you.

4 Make sure the two cogs are touching each other and that the corrugated ribs interlock.

5 Decorate the box with colored cardboard.

ADDING COGS

Add a third cog to your system of gears and look at what happens. You will find that the final wheel turns in the same direction as the first wheel.

6 Turn one cog. The other cog turns as well, but in the opposite direction. Count how many times the second cog turns for every turn of the first cog. Try cogs of different sizes and see how this affects the number of turns of each one.

*T*YPES OF GEARS

THE LAST PROJECT SHOWED how cogs work. There are many other types of gears that are used to transmit power in a machine. Many of these are wheel-shaped, like cogs, but some can be flat, while others can involve chains or even belts. Build a multi-geared machine in this project and see how gears transmit power from one place to another. Ask an adult to help you with all the steps.

WHAT YOU NEED

Cardboard
Corrugated cardboard
Glue
Cardboard box
Wide and thin drinking straws
Matchsticks

GEARING UP

1 Cut slots into the lid of the cardboard box as shown below.

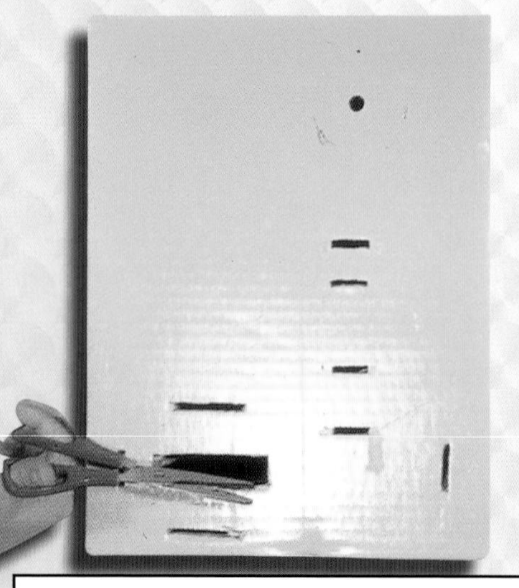

2 Using the method shown on page 26, make three cogs.

3 Cut out two disks of cardboard. Glue upright matchsticks around the edges (A). Make a card frame (B). Push a straw through the frame. Push a disk and a cog onto each end.

4 Cut a thin strip of corrugated cardboard. Glue it to one side of a wide straw. On the other side, glue on a strip of cardboard with teeth cut into it (C). Cut out two cards with holes in them (D).

(D)

(B)

(C)

(A)

(E)

(G)

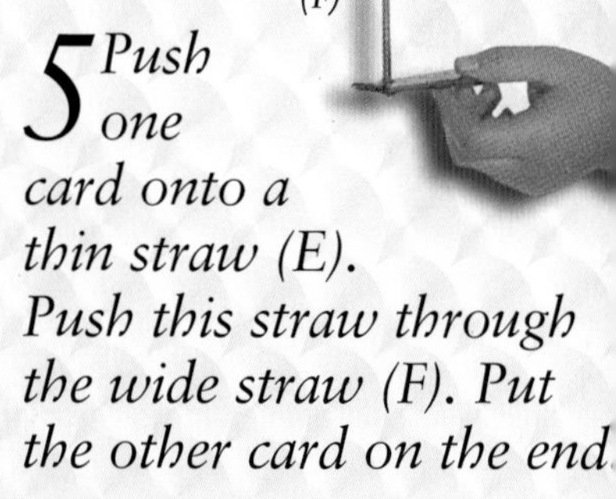

5 Push one card onto a thin straw (E). Push this straw through the wide straw (F). Put the other card on the end.

6 Cut out a star cog. Push a straw through it and put a card onto each end (G).

MORE GEARS

Look around and see where you can find various types of gears in action. What about on your bike?

WHY IT WORKS

Force is transmitted through gears in several ways. A rack and pinion system (1) uses a flat set of teeth against a cog. Bevel gears (2) lie at angles to each other and transmit their force at these angles. Spur gears (3) involve two flat cogs, like the ones in the last project.

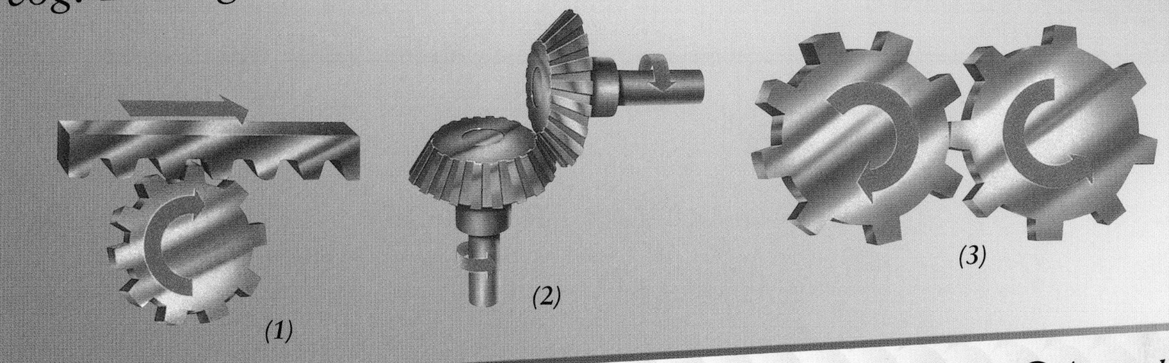

7 Join the other disks to a cog with a short straw.

8 Assemble all the pieces onto the slits cut into the lid of the box, as shown here. Turn the first cog and watch as the other gears move.

GLOSSARY

ABILITY (uh-BIH-luh-tee) The power to do something. *The project on pages 18-19 shows you some different abilities of wheels.*

AXLE (AK-sul) This is a bar that connects a number of wheels. *The project on pages 16-17 shows you how an axle connected to a steering wheel can turn a car.*

ENERGY (EH-nur-jee) The power to work or to act. *The project on pages 14-15 shows you how friction increases the energy needed to do something.*

FLYWHEEL (FLY-hweel) This is a heavy wheel that stores energy as it spins. This means it can keep spinning for a long time. *Find out where you can see a flywheel in action on pages 18-19.*

FRICTION (FRIK-shin) This is the force that is created when two objects rub against each other. The force of friction acts against movement and can slow objects down. *You can see how friction can be helpful in climbing a steep slope in the project on pages 12-13.*

GEARS (GEERZ) These are objects that are used to transmit force (move force from one place to another). Gears can be wheels with teeth, flat bars with teeth, or even screws. *Find out about different types of gears, and what they are used for, on pages 26-27 and 28-29.*

Glossary

LEVER (LEH-vur) This is a simple machine, usually a solid bar, which is used to transmit a force around a pivot to a load. *You can build your own levers and see all the different types of levers in the projects on pages 6-7 and 8-9.*

PIVOT (PIH-vut) This is a point about which an object rotates (moves in a circle). *You can see how pivots help levers to move objects on pages 6-7 and 8-9.*

PULLEY (PU-lee) This is a wheel around which a rope is pulled to transmit force. Pulleys working together can reduce the work level (the amount of effort you put in) needed to lift an object. *You can find out how to make your own pulley in the project on pages 22-23.*

RAMP (RAMP) A sloping platform. *Find out how friction on a ramp can be reduced by using different materials on pages 14-15.*

SURFACES (SER-fes-ez) The outsides of things. *The project on pages 16-17 shows you how wheels reduce the friction between surfaces.*

WINCH (WINCH) This is a hand-turned wheel that can be used to drive a machine or to wind in a rope. *You can see how a winch can lift a bucket in the project on pages 20-21.*

INDEX